Special Digest 1

Concrete in aggressive ground

Part 2: Specifying concrete and additional protective measures

BRE Centre for Concrete Construction
BRE Centre for Ground Engineering and Remediation

BRE
constructing the future

dti
Department of Trade and Industry

Prices for all available BRE publications can be obtained from:
CRC Ltd
151 Rosebery Avenue
London, EC1R 4GB
Tel: 020 7505 6622
Fax: 020 7505 6606
email:
crc@construct.emap.co.uk

SD1 Part 2
ISBN 1 86081 503 0

First published 2001

Published by
Construction Research Communications Ltd
by permission of
Building Research Establishment Ltd

The four parts of this Special Digest deal with the specification of new concrete that will be used in the UK in contact with aggressive ground.

The Special Digest applies both to buildings and civil engineering construction. It does not cover the diagnosis, remedy or structural implications of chemical attack, nor does it cover safety and health aspects of aggressive ground.

It replaces Digest 363, which is now withdrawn.

Acknowledgements
The preparation of this Special Digest was funded by a Government sponsored construction research programme. The work was undertaken in close collaboration with a steering committee comprising members of the BS 5328 Thaumasite Amendment Panel. Contributions made by numerous consultees and, in particular, Members of the Thaumasite Expert Group, are acknowledged.

Contents

Parts 1 to 4 of this Special Digest

Part 1 identifies chemicals in the ground that are potentially harmful to concrete and gives procedures that lead to assessment of the Aggressive Chemical Environment for Concrete (ACEC) Class of the ground.

Part 2 provides recommendations for the types of cement and aggregate and the type and quality of concrete suitable for use below ground level in the aggressive chemical environments identified and classified in Part 1. Concrete quality is specified in terms of a Design Chemical Class (DC Class) corresponding to each combination of ACEC Class and use of concrete below ground. Each DC Class (Table 6) follows the previous practice of defining concrete quality in terms of minimum cement content and maximum water/cement ratio for each cement group. However, it also allows for any carbonate content of the aggregates used in the concrete by additionally specifying concrete in terms of Aggregate Carbonate Range. This latter measure has been introduced to combat the thaumasite form of sulfate attack (TSA), following the recommendations of the Thaumasite Expert Group.

A procedure, which broadly follows that of previous Digests, is given in Table 7 for deriving concrete specification in terms of DC Class. The procedure either allows a relaxation or recommends an increase in the concrete quality according to the thickness of the concrete element and the conditions of exposure (eg exposure to hydrostatic groundwater pressure). In some circumstances, Additional Protective Measures (APM) are recommended to further protect the concrete. The various APM options are listed in Table 8. The number of APM recommended depends on the Structural Performance Level required of the concrete element, as defined in Table 5.

Part 3 gives sets of Design Guides 1x to 3x for the common applications of below-ground concrete, including foundations, retaining walls and floor slabs for domestic properties. These Design Guides inherently allow for the variation in type of concrete construction (eg concrete section thickness) or exposure conditions (eg hydrostatic pressure and carbonation). These parameters are taken into account by adjustment of the relationship between the ACEC Class for the ground and the DC Class for the concrete and, in some cases, by recommending specific Additional Protective Measures. They also carry specific design notes based on expert experience. It is therefore recommended that Design Guides 1x to 3x be used and be referenced in contract documents, wherever they are applicable.

Part 4 gives sets of Design Guides 4x to 7x for the specific precast concrete products used for pipeline systems, segmental linings for tunnels and shafts, box culverts and concrete masonry blocks. These Design Guides allow for the exposure conditions likely to be encountered by these products. They also take into account distinctive concrete specifications arising from particular methods of manufacture.

The Tables are numbered consecutively irrespective of the Part in which they appear

1. Introduction

This part of this Special Digest provides guidance on the progression from the ACEC Class of the ground, derived in Part 1 (restated here as Table 2), into the requirements for the structure, namely the necessary concrete quality and the number of Additional Protective Measures.

The procedure given here for arriving at a concrete specification broadly follows that of previous BRE Digests in taking account both of ground conditions and the form and use of the concrete element. For each ACEC Class, concrete quality is specified (in Table 7) in terms of a Design Chemical Class (DC Class), taking account of Structural Performance Level (SPL), concrete thickness and the conditions of exposure (for example exposure to hydrostatic pressure).

Each DC Class in Table 6 follows the previous practice of defining concrete quality in terms of minimum cement content and maximum free water/cement ratio for each cement group. However, it also allows for any carbonate content of the aggregates used in the concrete by additionally specifying concrete in terms of Aggregate Carbonate Range. This latter measure has been introduced to combat the thaumasite form of sulfate attack (TSA), following the recommendations of the Thaumasite Expert Group [1].

In some cases, Additional Protective Measures (APM) are recommended in Table 7 to further protect the concrete. The number of APM needed increases with higher ACEC Class of the ground and with higher Structural Performance Level required for the concrete element. The various APM options are listed in Table 8.

The generic procedure for deriving the DC Class and accompanying APM based on Tables 6 and 8 can be complex for some types of construction. An alternative user-friendly procedure is, therefore, provided using Design guides for common applications of concrete and for specific precast concrete products in Parts 3 and 4 of this Special Digest (see Appendix 6).

The overall process of design of concrete for use in aggressive ground conditions is shown diagrammatically in Figure 2-1. The figure indicates the scope of the four parts of this Special Digest. A summary of the procedure given in this part for the derivation of DC Class and APM is set out in Figure 2-2. A glossary of terms used in this Digest is in Appendix 5 on page 18.

2. Concrete materials

2.1 Cement Type

The groups of cements and combinations recommended for use in aggressive ground are listed in Table 3. These groups follow traditional current British Standard nomenclature, but a future revision of this Special Digest will cater for emerging European Standards, which will displace some British Standards (see Section 7 and Appendix 2).

The groups in Table 3 follow previous BRE Digests in being based primarily on their record of performance in resisting the ettringite and gypsum form of sulfate attack. Precautions to combat TSA have been additionally introduced in Table 6 in the form of higher minimum cement contents and lower maximum free water/cement ratios for the concretes and by specification in terms of Aggregate Carbonate Range. Research into the contribution that various types of cement make to the ability of concrete to resist TSA is currently in progress.

No restrictions on the type of cement to resist acid attack are given because the rate of erosion of concrete surfaces by natural acidic waters is affected less by the type of cement than by the quality of the concrete.

The expert use of special cements, such as supersulfated cement conforming to BS 4248 and calcium aluminate cement conforming to BS 915, can produce concrete with very good chemical resistance. These are discussed further in Appendix 1 on page 14.

In the case of magnesium sulfate, there is some evidence from laboratory tests that certain cements, in particular those containing ground granulated blastfurnace slag (ggbs) or pulverized-fuel ash (pfa), are more susceptible to attack by high concentrations of magnesium sulfate than is sulfate-resisting Portland cement. This is catered for by specific ACEC Classes AC 4m and AC-5m in Table 2 in Part 1 for magnesium concentrations greater than 1.2 g/l for soil/water extract and greater than 1.0 g/l for groundwater (but see Note [1] of the Table and Part 1, Section 2.6 in respect of chlorides).

Where ggbs-containing cements are used in concrete together with good quality carbonate aggregates, the susceptibility to attack by magnesium sulfate seems to be counteracted.

2.2 Aggregate Type

Previous Digests dealing with the chemical resistance of concrete have found no necessity to take the type of aggregate into consideration except with regard to limestone in some conditions of acid and aggressive CO_2 (see Part 1). However, the investigation of recent occurrences of TSA in below-ground concrete [1] has shown that the carbonate level in the aggregate is a contributory factor to this form of sulfate attack. Affected concretes have included coarse and fine aggregates containing crushed limestone or dolomite or limestone in sands and gravels. Laboratory tests to date have shown that both the coarse and fine aggregate contribute to the reaction. Moreover, the threshold for calcium carbonate ($CaCO_3$) equivalent at which TSA occurs is lower in the fine aggregate than in the coarse aggregate.

Three Ranges of aggregate combinations have accordingly been defined for use in this Special Digest, denoted A, B and C. The permitted quantities of carbonate in these (expressed as calcium carbonate equivalent, which caters also for dolomite) is given in Table 4. The limits given in the table are for particular distributions of carbonate in the coarse and fine fractions of the aggregate. Precise definition and procedures for determination of Aggregate Carbonate Range in any

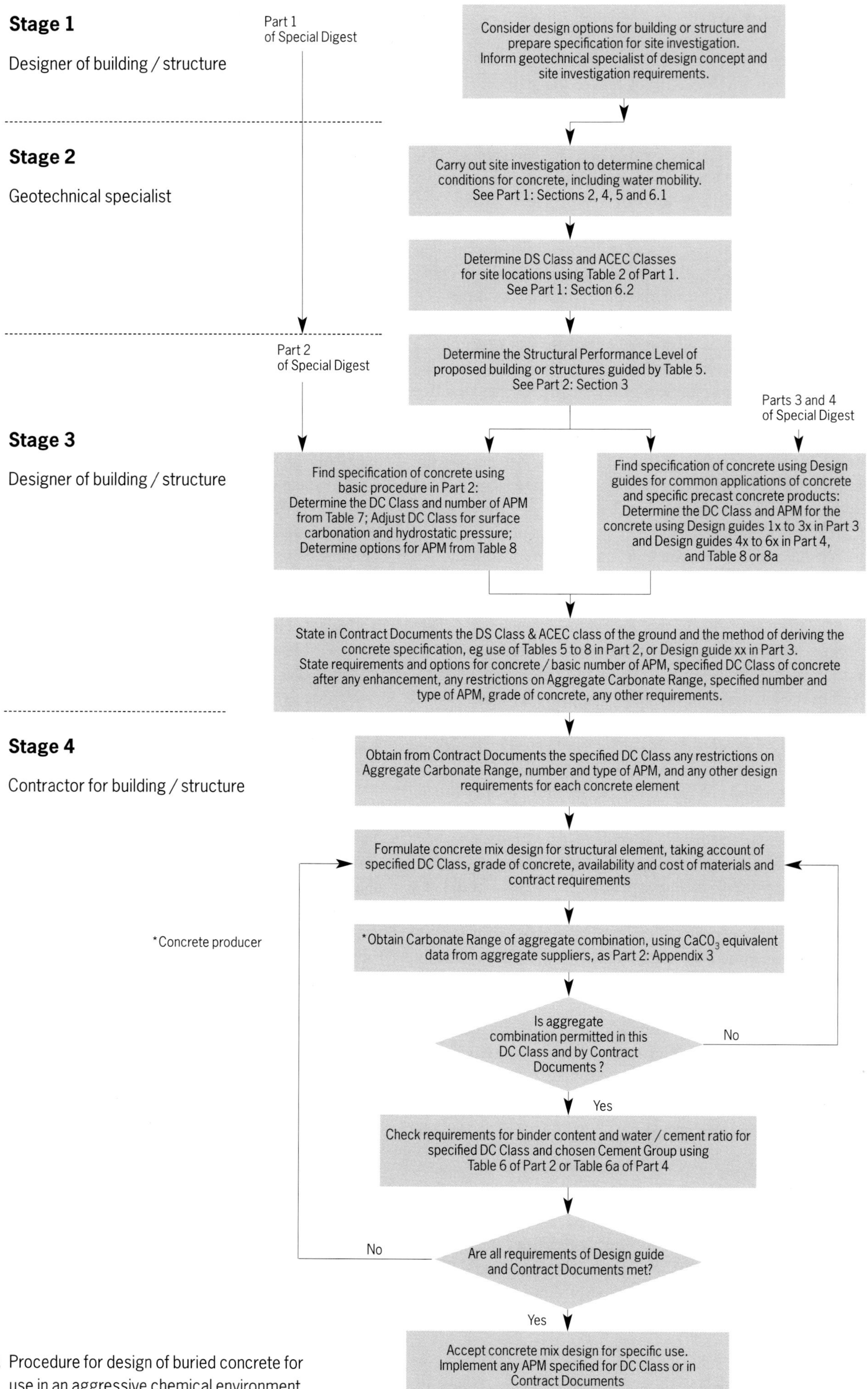

Figure 2-1 Procedure for design of buried concrete for use in an aggressive chemical environment

Table 2 Aggressive Chemical Environment for Concrete (ACEC) site classification
Refer to Part 1, Sections 4, 5 and 6 before using this table for ground assessment

Sulfate and magnesium						Natural soil		Brownfield [3]		ACEC class for site
Design Sulfate Class for site	2:1 water/soil extract		Groundwater		Total Potential sulfate [2]	Static water	Mobile water	Static water	Mobile water	
1	2	3	4	5	6	7	8	9	10	11
	SO_4 g/l	Mg g/l	SO_4 g/l	Mg g/l	SO_4 %	pH	pH	pH [4]	pH [4]	
DS-1	<1.2		<0.4		<0.24	All pH values		All pH values		AC-1s
							>5.5		>6.5	AC-1
							≤5.5		5.5–6.5	AC-2z
									4.5–5.5	AC-3z
									<4.5	AC-4z
DS-2	1.2–2.3		0.4–1.4		0.24–0.6	>3.5		>5.5		AC-1s
							>5.5		>6.5	AC-2
						≤3.5		≤5.5		AC-2s
							≤5.5		5.5–6.5	AC-3z
									4.5–5.5	AC-4z
									<4.5	AC-5z
DS-3	2.4–3.7		1.5–3.0		0.7–1.2	>3.5		>5.5		AC-2s
							>5.5		>6.5	AC-3
						≤3.5		≤5.5		AC-3s
							≤5.5		5.5–6.5	AC-4
									<5.5	AC-5
DS-4	3.8–6.7	≤1.2	3.1–6.0	≤1.0	1.3–2.4	>3.5		>5.5		AC-3s
							>5.5		>6.5	AC-4
						≤3.5		≤5.5		AC-4s
							≤5.5		≤6.5	AC-5
DS-4m	3.8–6.7	>1.2 [1]	3.1–6.0	>1.0 [1]	1.3–2.4	>3.5		>5.5		AC-3s
							>5.5		>6.5	AC-4m
						≤3.5		≤5.5		AC-4ms
							≤5.5		≤6.5	AC-5m
DS-5	>6.7	≤1.2	>6.0	≤1.0	>2.4	>3.5		>5.5		AC-4s
						≤3.5	All pH values	≤5.5	All pH values	AC-5
DS-5m	>6.7	>1.2 [1]	>6.0	>1.0 [1]	>2.4	>3.5		>5.5		AC-4ms
						≤3.5	All pH values	≤5.5	All pH values	AC-5m

Notes

[1] The limit on water-soluble magnesium does not apply to brackish groundwater (chloride content between 12g/l and 18g/l). This allows **m** to be omitted from the relevant ACEC classification.

[2] Applies only to sites where concrete will be exposed to sulfate ions (SO_4) which may result from the oxidation of sulfides such as pyrite, following ground disturbance.

[3] 'Brownfield' is defined as sites which may contain chemical wastes remaining from previous industrial use or from imported wastes.

[4] An additional account is taken of hydrochloric and nitric acids by adjustment to sulfate content - see Section 6.1.2.

Explanation of suffix symbols to ACEC Class number

- Suffix **s** indicates that, as the water has been classified as Static, no Additional Protective Measures are generally necessary.
- Concrete placed in ACEC Classes which include the suffix **z** have primarily to resist acid conditions and may be made with cements from any of the Groups in Part 2, Table 3 – but see the restrictions in the table on the use of Portland limestone cement.
- Suffix **m** relates to the higher levels of magnesium in Sulfate Classes 4 and 5.

given aggregate combination is addressed in Appendix 3 on page 16. Because aggregate fractions may derive from different sources, it is generally the concrete producer who ultimately determines the Aggregate Carbonate Range of an aggregate combination using data from the aggregate producers on the percentage of $CaCO_3$ equivalent in the coarse and fine fractions.

Where Range B or C aggregates are being considered as a specification requirement, check their availability with local concrete suppliers. Where they are not normally stocked, give sufficient notice to the producer so that they can source supplies and develop the appropriate mix designs.

3. Structural performance level (SPL)

In previous Digests, recommendations for the durability of foundation concrete have tended to be influenced by various structural performance factors, such as the consequence of serious erosion and ease of repair. However, these concepts were not explicitly included. In this Special Digest, the concept of Structural Performance Level (SPL) has been introduced specifically to take account of such factors. Three Levels have been defined: Low, Normal and High. The typical attributes of each Level are given in Table 5.

The Design guides given in Parts 3 and 4 of this Special Digest take account of the anticipated SPL appropriate to different categories of structure such as a 'Low' level for a domestic building. It is, however, open to clients and designers to decide the SPL they require, for example, a 'High' level for some specific purpose.

Table 3 Cement or combination groups for use in Table 6

Group	Description
1	a) Portland cement conforming to BS 12
	b) Portland blastfurnace cements conforming to BS 146
	c) High slag blastfurnace cement conforming to BS 4246
	d) Portland pulverized-fuel ash cements conforming to BS 6588
	e) Pozzolanic pulverized-fuel ash cement conforming to BS 6610
	f) Portland limestone cement conforming to BS 7583
	g) Combinations of Portland cement conforming to BS 12 with ggbs conforming to BS 6699
	h) Combinations of Portland cement conforming to BS 12 with pulverized-fuel ash conforming to BS 3892 : Part 1
2	a) Portland pulverized-fuel ash cements conforming to BS 6588 containing not less than 26% of pfa by mass of the nucleus [1] or combination of Portland cement conforming to BS 12 with pfa conforming to BS 3892 : Part 1, where there is not less than 25% pfa and not more than 40% pfa by mass of the combination
	b) High slag blastfurnace cement conforming to BS 4246, containing not less than 74% slag by mass of nucleus [1] or combinations of Portland cement conforming to BS 12 with ggbs conforming to BS 6699 where there is not less than 70% ggbs and not more than 85% ggbs by mass of the combination [2]
3	Sulfate-resisting Portland cement conforming to BS 4027

Notes

[1] The nucleus is the total mass of the cement constituents excluding calcium sulfate and any additives such as grinding aids.

[2] For all group 2b cements, granulated blastfurnace slag with alumina content greater than 14% should only be used with Portland cement having a tricalcium aluminate (C_3A) content not exceeding 10%.

Table 4 Aggregate Carbonate Range limits for various aggregate combinations, expressed as calcium carbonate equivalent as a percentage of the total aggregate mass

	Aggregate Carbonate Range [1]		
	Range A **% $CaCO_3$ Equiv.**	**Range B** **% $CaCO_3$ Equiv.**	**Range C** **% $CaCO_3$ Equiv.**
All the carbonate in the coarse fraction	Min = 30 Max = 108 [2]	Min = 12 Max = 30	Min = 0 Max =12
All the carbonate in the fine fraction	Min = 10 Max =108 [2]	Min = 2 Max = 10	Min = 0 Max = 2
Carbonate in both coarse and fine fraction see Figure A2.1	Min = 10 to 30 Max =108 [2]	Min = 2 to 12 Max = 10 to 30	Min = 0 Max = 2 to 12

Notes

[1] See Appendix 3 for methods of determination and calculation

[2] For calcium carbonate the upper limit is 100% and for dolomite it is 108% – see Appendix 3

4. Composition of concrete to resist chemical attack

4.1 Design Chemical Class(DC Class)

This is a new system for defining 'qualities' of concrete that are required to resist chemical attack. The DC Class is derived from the ACEC Class of the ground (Table 2) taking into account a number of factors, including the type of concrete element, the method of curing, its mode of exposure to the aggressive ground, and the required Structural Performance Level. Procedures for the derivation of DC Class are discussed in Section 4.2.

For a given DC Class, specifications for concrete are given in Table 6. The table gives options so that, in most cases, the concrete producer can use the aggregates and cements that are normally stocked. Where specification needs to restrict the Aggregate Carbonate Range (ACR), this is done by specifying one of the starred (*) classes. The required quality of concrete is then assured by specifying a minimum cement content and maximum free water/cement ratio.

The requirements for composition of concretes have been made more stringent, compared with previous Digests, to provide better resistance to TSA. It is now necessary to categorise the combined fine and coarse aggregate with respect to their carbonate content. Where Range A aggregates are used in Design Class DC-3 and above, an increase in minimum cement content and reduction in maximum free water/cement ratio is recommended to provide 'basic' resistance to TSA. Where such an increase in minimum cement content and reduction in maximum free water/cement is combined with the use of Range B or Range C aggregates, enhanced resistance to TSA is provided. Such fortified DC Classes are indicated in Table 6 by single and double stars respectively. Their optional use to offset the recommended number of APM is explained in Notes [7] and [8] of Table 6 and Note [5] of Table 7.

The effectiveness of these concretes to resist chemical attack depends to a high degree on their impermeability. Therefore, good compaction is most important. With low water/cement ratios, such as those advocated here, it is probable that water reducing admixtures will be necessary to achieve effective compaction. This will be particularly true of concretes such as those used in piling where mechanical compaction cannot be used.

4.2 Determining DC Class for specification of concrete

4.2.1 Two procedures for determining DC Class

Two procedures are given in this Special Digest for determining the appropriate DC Class of concrete for given ground conditions and use of concrete:

(i) A generic procedure based on Table 7 which gives a range of specifications for all concrete to be used in aggressive ground, together with the number and type of Additional Protective Measures. The use of Table 7 is explained in Section 4.2.3. A comparison with the procedure for concrete specification given in previous Digest 363 is included as Appendix 4. For some cases, specification using this generic procedure requires expert knowledge of the possible processes of chemical attack and of the attributes of concrete that offer chemical resistance.

(ii) Because of the difficulty the non-expert may experience using Table 7, a second procedure is given in Parts 3 and 4 based on the use of sets of Design guides 1x to 7x for particular types of concrete element. These Design guides inherently allow for variation in type of concrete construction (for example, concrete section thickness) or exposure conditions (for example, hydrostatic pressure and carbonation). The variations are accommodated by adjustment of the relationship between the ACEC Class or Design Sulfate Class for the ground and the DC Class for the concrete, and, in some cases, by recommending specific Additional Protective Measures. They also carry specific design notes based on expert experience. It is recommended, therefore, that Design guides 1x to 7x of Parts 3 and 4 are used, and referenced in contract documents, wherever they are applicable.

Table 5 Structural Performance Level of buried concrete

Structural Performance Level (SPL)	Typical attributes
Low	Short service life structures less than 30 years; Unreinforced concrete; Non-critical structural details; Temporary structures; Long service life structures, but with associated low stress levels, eg house foundations (unreinforced).
Normal	Intermediate service life (30 to 100 years); Not falling in either high or low category.
High	Long service life structures greater than 100 years, eg, transport structure foundations; Vulnerable critical details such as slender structural elements, hinges, joints; Structures retaining hazardous materials.

The general principles used in compiling the DC Classes and number and type of APM given in Table 7 and Design guides 1x to 7x are set out in Sections 4.2.2 and 5.2 respectively. In some cases, specifications that would have been predicted by use of these principles have been modified to accord with expert engineering judgement. Since the identical principles have been followed in composing sets of Design guides 1x to 3x and Table 7, the former may be used as a model for closely related generic applications based on Table 7. Some amended principles, however, apply to Specific Precast Concrete Products (Design guides 4x to 7x) as explained in Part 4.

4.2.2 Principles used to determine DC Class

The general principles used as a basis for determining DC Classes given in Table 7 and Design guides 1x to 7x are:

(i) For cast-in-situ concrete elements thicker than 450 mm, the DC Class has been reduced by one relative to concrete of 'basic' thickness, (140 – 450 mm) provided durability and structural conditions permit. Applying this class adjustment implies that some small degree of sulfate attack affecting the surface of the concrete is acceptable. Using this adjustment is therefore inappropriate where the surface of the concrete is required to remain intact, for example in elements such as friction piles.

(ii) For concrete which is surface carbonated, the DC Class may be reduced by one relative to the DC Class for normal uncarbonated concrete. Alternatively the number of any APM may be relaxed by 1 (Note [1] to Table 7). Such concrete may include precast ground beams, wall units and piles, where these are normally cured, with the surface of concrete carbonated. This reduction, associated with carbonation of the concrete surface, should not be made if the ACEC Class has been derived from a low pH in Design Sulfate Classes 1 and 2 (suffix z) as surface carbonation does not protect against acid attack.

(iii) For thin section concretes, less than 140 mm thick, either, the recommended DC Class has been increased by one relative to concrete of 'basic' thickness (140 - 450 mm) or, the use of an additional APM has been recommended. A section thickness of less than 140 mm is generally not recommended for concrete intended for use in High Structural Performance Level elements.

(iv) Where the concrete will be subject to a hydrostatic head of groundwater greater than five times the section thickness, the DC class may remain the same as for concrete not so exposed, but it is recommended that one additional APM is applied (see (vi) in Section 5.2).

(v) Use of suffixes to DC Class:
If the originating ACEC Class carried a suffix 'm' or 'z' these should be similarly affixed to the resultant DC Classifications. Suffix notations 'z' indicate concretes that primarily must resist acid conditions and 'm' indicate concretes which must resist high levels of magnesium sulfate. The concrete specification needs only these two suffix notations as suffix 's' is taken into account by dispensing with APM (with some exceptions) where site assessment has shown groundwater conditions to be static. For DC Classes of DC-3 and above, 'starred' Classes offer optional alternatives.

4.2.3 Using Table 7 to determine DC Class

Table 7 gives recommendations for concrete specification in terms of DC Class for each of the ACEC Classes defined in Table 2. Recommendations are for each of the appropriate three Structural Performance Levels (see Section 3) and, within these, two or three concrete thickness categories. The detailed procedure based on Table 7 for determining concrete specification and APM is set out diagrammatically in Figure 2-2.

5. Additional protective measures (APM)

5.1 Options and number of APM

There is little field data on which to build recommendations to resist TSA and laboratory tests are incomplete. Until more evidence is available, it was considered by the Thaumasite Expert Group that the basic provisions for concrete specification (Table 6) might not provide adequate resistance to attack in some instances and the use of Additional Protective Measures (APM) was recommended.

The number of these APM which are recommended for each ACEC Class, Structural Performance Level (SPL) and thickness of concrete section are shown in Table 7. More APM need to be simultaneously adopted as ground conditions incline to being more highly aggressive and/or a higher SPL is required. No APM are generally required where the ACEC Class has a suffix 's', indicating static ground water conditions, defined in Part 1. An exception is where the hydrostatic head across the concrete is more than five times the section thickness (see Note 2).

The five options for APM given in this Special Digest are listed in Table 8. For practical reasons each APM is given equal status. In most cases, this can be used to select any of the APM options, however, there are some specific requirements for DC-4, DC-4m and DC-4z concretes as indicated in Table 7 and explained in Section 5.2. The measures are described in Sections 5.3 to 5.7.

In addition to specified APM, in a few situations (for example in Table 7, for AC-5 and AC-5m conditions in conjunction with High SPL), this Special Digest additionally recommends the use of double-starred DC Classes of concrete which contain Range C aggregate.

In most other cases where APM are recommended, starred DC Classes can optionally be used to offset up to 2 of the required APM (see Notes [7] & [8] of Table 6 and Note [5] of Table 7).

Where APM are recommended, those relating to specification of the concrete will require liaison between the construction designer and the concrete producer.

5.2 Determining the APM to be used in concrete specification

As in the determination of DC Class (Section 4.2) there are two ways to determine the number and type of APM to be used as part of a concrete specification:

(i) The first route shown in Figure 2-2, is using Tables 7 and 8. From Table 7, the recommended number of APM is found attached to the DC Class previously selected for use (see Section 4.2.3). Appropriate options for APM can then be selected from Table 8 taking into account the constraints on choice of APM options listed in the Notes to Table 7.

(ii) The second and recommended route for common applications of below-ground concrete, is by using Design guides 1x to 7x in Parts 3 and 4. In these guides the last two columns give the recommended number of APM and, where relevant, recommend that a

Table 6 Concrete qualities to resist chemical attack

Design Chemical Class [5]	Aggregate Carbonate Range	Cement or combination group [1]	Dense fully compacted concrete made with aggregate conforming to BS 882 or BS 1047	
			Minimum cement content kg/m³	Maximum free water / cement ratio
DC-1	No restriction	1,2,3	-	-
DC-2	A[4], B, C	1[2]	340	0.50
	A[4], B, C	2, 3	300	0.55
DC-2z [3]	No restriction	1[2], 2, 3	300	0.55
DC-3	A	2a	400	0.40
	A	2b, 3	380	0.45
	B, C	2, 3	340	0.50
DC-3* [7]	B	2, 3	380	0.45
DC-3** [8]	C	2, 3	380	0.45
DC-3z [3]	No restriction	1[2], 2, 3	340	0.50
DC-4	A	2a	400	0.35
	A	2b, 3	400	0.40
	B, C	2, 3	380	0.45
DC-4* [7]	B	2, 3	400	0.40
DC-4** [8]	C	2, 3	400	0.40
DC-4z [3]	No restriction	1[2], 2, 3	380	0.45
DC-4m	A	2b [6], 3	400	0.40
	B, C	3	380	0.45
DC-4m * [7]	B	3	400	0.40
DC-4m** [8]	C	3	400	0.40

Notes

[1] For cement or combination groups see Table 3.

[2] Portland limestone cement may be used only in concrete specified as Design Chemical Classes DC-2, DC-2z, DC-3z, DC-4z where the sulfate classification (Design Sulfate class for the site - see Table 2) does not exceed DS-1.

[3] Classes DC-2z, DC-3z and DC-4z apply where chemical resistance is recommended primarily to resist acid attack under conditions described in this Special Digest.

[4] In addition to the requirement for a minimum cement content and a maximum free water/ cement ratio, a minimum concrete Grade of C35 is recommended when using Range A aggregate combinations.

[5] Bold horizontal lines indicate division of the Table in terms of concrete quality in respect of aggressive ground conditions. Using the option 'Enhance concrete quality' as an Additional Protective Measure can be satisfied by using the recommendations of the next numerically higher DC Class. Where the starting point is a 'starred' DC Class, the step is to a similarly starred higher DC Class, eg DC-3** plus a 1 step enhancement of concrete quality leads to a specification of DC-4**.

[6] Use of group 2b cements in DC-4m concrete requires use of Range A aggregate which has been strictly determined on the basis of carbonate content, and not by 'declaration' (see Part 2, Appendix 3).

[7] Using a single-starred DC Class may permit the recommended number of APM to be reduced by 1 (see Note [5] of Table 7).

[8] Using a double-starred DC Class may permit the recommended number of APM to be reduced by 2 (see Note [5] of Table 7).

First, determine parameters:
- ACEC Class of ground as Part 1, Section 6.2;
- Structural Performance Level of concrete element from Table 5;
- Thickness of concrete element (see categories in Table 7);
- Hydrostatic exposure conditions and state of carbonation of concrete element (see Notes [1] and [2] to Table 7).

↓

From Table 7, determine appropriate DC Class of concrete and number of APM:
- For the assessed ACEC Class, look in the column corresponding to the required Structural Performance Level and thickness of concrete section;
- Adjust DC Class or Number of APM to take account of assured surface carbonation as Note [1];
- Where hydrostatic pressure exceeds 5 times section thickness, apply an extra APM as Note [2].

↓

From Table 7:
- For AC-5 and AC-5m conditions, note any recommendation to use double-starred DC Classes which utilise only Range C aggregate;
- For DC Classes DC-3 and above, consider optional use of 'starred' DC Classes, eg DC-3** to reduce required number of APM.

↓

From Table 7, note any restrictions to choice of APM, eg use of any APM3.

↓

From Table 8, select appropriate options for APM, taking account of any restrictions.

↓

Include in the contract documents:
- Design Sulfate (DS) Class of ground;
- ACEC Class of ground;
- Any account taken of carbonation and/or hydrostatic conditions;
- Basic DC Class/number of APM, as boxes in Table 7;
- Specified DC Class after optional adjustment/enhancement;
- Any restrictions on Aggregate Carbonate Range;
- Specified number of APM after adjustment;
- Any restrictions/preferences in respect of APM to be used.

Continue as Stage 4 of Figure 2-1.

Figure 2-2 Derivation of DC Class and APM from Tables 7 and 8 in Part 2

Table 7 Concrete quality and number of Additional Protective Measures recommended for the general use of in-situ and precast concrete

	Design Chemical (DC) Class / Number of Additional Protective Measures (APM) [1] [2]							
	Low Structural Performance Level			**Normal Structural Performance Level**			**High Structural Performance Level** [3]	
Section thickness	<140 mm	140 - 450 mm	>450 mm [4]	<140 mm	**140 - 450 mm**	>450 mm [4]	140 - 450 mm	>450 mm [4]
ACEC class								
AC-1	DC-2/0	DC-1/0	DC-1/0	DC-2/0	**DC-1/0**	DC-1/0	DC-1/0	DC-1/0
AC-1s	DC-2/0	DC-1/0	DC-1/0	DC-2/0	**DC-1/0**	DC-1/0	DC-1/0	DC-1/0
AC-2	DC-3/0	DC-2/0	DC-1/0	DC-3/0	**DC-2/0**	DC-1/0	DC-2/0	DC-1/0
AC-2s	DC-3/0	DC-2/0	DC-1/0	DC-3/0	**DC-2/0**	DC-1/0	DC-2/0	DC-1/0
AC-2z	DC-3z/0	DC-2z/0	DC-1/0	DC-3z/0	**DC-2z/0**	DC-1/0	DC-2z/0	DC-1/0
AC-3	DC-3/2 [5]	DC-3/1 [5]	DC-2/1	DC-3/3 [5]	**DC-3/2** [5]	DC-2/2	DC-3/3 [5]	DC-2/3
AC-3s	DC-4/0	DC-3/0	DC-2/0	DC-4/0	**DC-3/0**	DC-2/0	DC-3/0	DC-2/0
AC-3z	DC-4z/0	DC-3z/0	DC-2z/0	DC-4z/0	**DC-3z/0**	DC-2z/0	DC-3z/0	DC-2z/0
AC-4	DC-4/2 [5] [6]	DC-4/1 [5]	DC-3/1 [5]	DC-4/3 [5]	**DC-4/2** [5]	DC-3/2 [5]	DC-4/3 [5]	DC-3/3 [5]
AC-4s	DC-4/0	DC-4/0	DC-3/0	DC-4/0	**DC-4/0**	DC-3/0	DC-4/0	DC-3/0
AC-4z	DC-4z/1 [7]	DC-4z/0	DC-3z/0	DC-4z/1 [7]	**DC-4z/0**	DC-3z/0	DC-4z/0	DC-3z/0
AC-4m	DC-4m/2 [5] [6]	DC-4m/1 [5]	DC-4m/0	DC-4m/3 [5]	**DC-4m/2** [5]	DC-4m/1 [5]	DC-4m/3 [5]	DC-4m/2 [5]
AC-4ms	DC-4m/0	DC-4m/0	DC-3/0	DC-4m/0	**DC-4m/0**	DC-3/0	DC-4m/0	DC-3/0
AC-5	DC-4/1 [7]	DC-4/1 [7]	DC-3/1 [7]	DC-4/3 [5] [8]	**DC-4/2** [5] [8]	DC-3/2 [5] [8]	DC-4**/1 [7]	DC-3**/1 [7]
AC-5m	DC-4m/1 [7]	DC-4m/1 [7]	DC-4m/1 [7]	DC-4m/3 [5] [8]	**DC-4m/2** [5] [8]	DC-4m/1 [7]	DC-4m**/1 [7]	DC-4m**/1 [7]
AC-5z	DC-4z/1 [7]	DC-4z/1 [7]	DC-3z/1 [7]	DC-4z/1 [7]	**DC-4z/1** [7]	DC-3z/1 [7]	DC-4z/1 [7]	DC-4z/1 [7]

Notes

[1] Where carbonation of the concrete prior to exposure can be assured, the recommended DC Class/APM (other than those carrying a z suffix to the DC Class), may be relaxed by either 1 DC Class or 1 APM, provided that for AC-5 and AC-5m conditions the resulting package includes APM3.

[2] Where the hydrostatic head of groundwater is greater than 5 times the section thickness one APM over and above the number indicated in the Table should be applied. This should preferably be APM5 but, if this is not practicable, apply either APM1 or APM3.

[3] For High SPL, a Section thickness of less than 140 mm is generally not recommended in aggressive ground conditions. Specific precast concrete products (see Part 4) are an exception.

[4] This column of the Table may be used for concrete elements greater than 450 mm thick where other durability and structural considerations permit. Use of this column is inappropriate where the surface of the concrete is required to remain unaffected by aggressive ground, for example in elements such as friction piles.

[5] Where DC-3, DC-4 or DC-4m is given, the number of APM may be reduced (provided that the reduction does not override the recommendation to use or include APM3) by:

- one provided that DC-3*, DC-4* or DC-4m* concrete is specified;
- two provided that DC-3**, DC-4** or DC-4m** concrete is specified.

If, in addition, the concrete quality is to be enhanced as a result of the application of APM1, the step in concrete quality is to the next higher * class, eg DC-4* is specified instead of DC-3* and DC-4** is specified instead of DC-3**.

[6] If APM3 is selected for Low SPL concrete no further APM are necessary.

[7] Only APM3 is recommended (not applicable to bored piles).

[8] To include APM3, where practicable, as one of the recommended number of APM.

Table 8 Options available to provide Additional Protective Measures for buried concrete

Option Code	Additional Protective Measure (APM) [1]
APM1	Enhanced concrete quality (see Section 5.3)
APM2	Use of controlled permeability formwork (see Section 5.4)
APM3	Provide surface protection (see Section 5.5)
APM4	Provide sacrificial layer (see Section 5.6)
APM5	Address drainage of site (see Section 5.7)

Note

[1] For options applicable to specific precast concrete products, see Part 4

specific APM should be used for some of the DC Classes, or that restrictions may be attached to the use of certain APM. In many cases, Notes attached to the Design guides explain the reasons for these limitations. Further explanation about reference in the Design guides to APM3 (provide surface protection) is given below.

The number of APM recommended in Table 7 and the Design guides are based on engineering judgement but accord with the following principles:

(i) The recommended number of APM increases with higher ACEC Class and higher Structural Performance Level (SPL) and decreases with a greater thickness of concrete section.

(ii) The recommendations are broadly in line with the Report of the Thaumasite Expert Group [1].

(iii) APM are generally not recommended in cases where the starting-point ACEC class had a suffix 's', indicating Static ground water conditions.

(iv) APM 3 'Provide surface protection' is recommended, where practicable, when using concretes that are required to combat Class AC-5, AC-5m or AC-5z aggressive ground conditions.

(v) For thin-section (<140 mm) concrete elements, the use of one extra APM is generally recommended when using DC-4, DC-4m and DC-4z concretes, since there is no higher concrete quality available.

(vi) Where the concrete will be subject to a hydrostatic head of groundwater greater than five times the wall thickness, the DC class may remain the same as for concrete not so exposed, but it is recommended that one additional APM is applied. Preferably this should be APM5 'Address drainage of site' to lower the hydrostatic pressure. Otherwise APM1 or APM3 is recommended (see Note [2] to Table 7).

(vii) Amended principles apply to Specific Precast Concrete Products (Design guides 4x to 7x) as explained in Part 4.

(viii) An accompanying restriction requiring use of Class DC-3** and DC-4** concrete with only Range C aggregates is recommended when both of the following apply:

❒ Concretes are required to combat AC-5 or AC-5m aggressive chemical conditions.

❒ The required SPL is 'High'.

5.3 Enhanced concrete quality - APM1

If a Design Chemical Class of DC-2 or DC-3 is derived by use of generic rules or from a design guide and Additional Protective Measures are also recommended, one Measure can be satisfied by increasing the concrete quality requirements by one DC Class, as delineated by the bold horizontal lines of Table 6. This option is not open to Classes DC-4 and DC-4m. For Class DC-2 the option can be applied twice over. If assessed ground conditions are AC-4m or AC-5m, any enhancement should be to a concrete quality specified in Table 6 which includes a suffix 'm' to the DC-Class. Similarly, enhancement by 1 DC Class from DC-3* or DC-3** calls for specification of DC-4* or DC4** respectively.

An example of the application of enhanced concrete quality is given in the box below.

Example of use of enhanced concrete quality as an APM

- Site assessed from Table 2 as AC-3.
- Concrete is to be used in a structural element of greater than 450 mm minimum section thickness, not subject to a hydrostatic head, and which is required to provide a High SPL.
- Table 7 recommends a concrete quality of DC-2/3, ie a basic DC-2 class concrete with three APM.

Option 1

Enhance concrete quality by 1 DC Class:
Use APM-1 once + two other APM. The concrete quality to be specified is DC-3 with any convenient aggregate.

Option 2

Enhance concrete quality by 2 DC Classes:
Use APM-1 twice + one other APM. The concrete quality to be specified is DC-4 with any convenient aggregate.

Option 3

Enhance concrete quality by 1 DC Class and use a one-star DC Class:
Use APM-1 once + one-star DC Class +one other APM. The concrete quality to be specified is DC-3* (has Range B aggregate).

Option 4

Enhance concrete quality by 1 DC Class and use a two-star DC Class:
Use APM-1 once + two-star DC Class (No additional APM now needed). The concrete quality to be specified is DC-3** (has Range C aggregate).

Option 5

Enhance concrete quality by 2 DC Class and use a one-star DC Class:
Use APM-1 twice + one-star DC Class (No additional APM now needed). The concrete quality to be specified is DC-4* (has Range B aggregate).

5.4 Use of controlled permeability formwork - APM2

The use of controlled permeability formwork (CPF) enhances the in-situ quality of the concrete in the cover zone relative to that achieved with conventional methods. It has been shown [2] to be able to produce a reduction in the water/cement ratio of concrete close to the interface with the formwork, extending to a depth of 10-15 mm into the concrete. Concomitant modifications of porosity have also been reported which, combined with the reduction in water/cement ratio, produce a very dense, low-porosity surface zone in concrete cast against CPF. Tests on this surface zone have indicated improvements in many of its properties compared with concrete cast against conventional formwork. These include improvements to durability-related properties such as permeability to water and oxygen, carbonation, freeze-thaw resistance and chloride ingress. Although no comparative testing of sulfate resistance has been reported, the above improvements in durability properties strongly indicates that sulfate resistance will be likewise enhanced.

The use of CPF should follow the manufacturer's recommendations.

5.5 Surface protection - APM3

Protection of concrete surfaces was recommended in previous Digests when the level of aggression arising from chemicals in the ground exceeded the maximum that the quality of concrete produced on site was considered able to resist. In the former Class 5 sulfate conditions, where no upper limit to the sulfate level was stated, the concrete quality recommended in Class 4 sulfate conditions was retained and surface protection was regarded as providing a further level of resistance. Whilst this concept is retained here for those same conditions, surface protection has also been introduced as an APM for chemical resistance at lower levels of SO_4. Two types of surface protection are considered here: coatings and water-resisting barriers.

5.5.1 Coatings

The main requirements of coatings are that they should:

- provide an impermeable barrier;
- be resistant to sulfates and other deleterious chemicals;
- have a neutral effect on the concrete substrate;
- be resistant to mechanical damage;
- be easy to apply;
- have long term durability;
- be cost effective.

In practice, the choice of coating will take account of the condition and accessibility of the surface and previous practical experience. Coatings have changed over the years, with tar and cut-back bitumens being less popular, so long-term field data on currently used materials are limited. Common current choices are rubberised bitumen emulsions. These should give good protection if well applied. Additionally, purpose-designed polymeric-based systems, for example epoxy resins, are now available. These coating systems can give exceptional performance, albeit at a higher initial cost.

The risk of damage to coatings during backfill operations should be considered.

Coatings must be applied in accordance with the manufacturer's instructions, and the workmanship must be of a high standard to maintain integrity.

5.5.2 Water-resisting barriers

The functional and practical requirements for water-resisting barriers are similar to those of coatings (see 5.5.1). Sheet materials are commonly used, including plastic and bituminous membranes. The former is commonly installed before placing the concrete: a 300 micron (1200 gauge) polyethythene membrane is commonly used to line excavations for trenchfill foundations in aggressive ground, or to cover a site prior to casting a raft foundation. Other types of membrane may be applied to the surface of the concrete after curing. The effectiveness of integral waterproofing agents in preventing sulfate attack is not established.

5.6 Sacrificial layer - APM4

The thickness of concrete is increased to absorb all the aggressive chemicals in a sacrificial outer layer. The quality of this additional concrete should be equal to or higher than that of the inner concrete. Using this measure is not appropriate where the surface of the concrete must remain sound to prevent loss of frictional resistance or settlement, for example for skin friction piles.

The life of the structure and the rate of penetration of chemicals into the concrete are the key issues that determine the required thickness of a sacrificial layer, but there is little guidance data. Field investigation of severe TSA on motorway bridge sub-structures, built with Portland cement concrete containing carbonate aggregate and buried in reworked pyritic clay, showed attack to a depth of up to 50mm in about 30 years [1]. The choice of Portland cement in this case was based upon the sulfate content of the pyritic clay during the original site investigation. However, subsequent backfilling with the clay appeared to have led to an increase in the sulfate content to a level for which the choice of Portland cement would have been inappropriate. See Section 6.1.3 of Part 1 for guidance on determination of potential sulfates due to oxidation of sulfides.

In using this example of the rate of penetration of TSA as a basis for recommending a suitable thickness for a sacrificial layer of concrete, it must be borne in mind that Portland cement was used, rather than one of the more suitable Group 2 or 3 cements listed in Table 3. The recommendations here should lead both to a more accurate assessment of aggressive ground conditions and to an appropriate specification for the concrete to be used. It seems reasonable, therefore, that together with other requisite APM, an additional surface protective layer of sacrificial concrete 50 mm thick would be adequate for, say, 120 years service life of a reinforced concrete structure.

This extra thickness of concrete should be treated as *additional to* the specified nominal cover, including situations where concrete is cast directly against the earth and the specified nominal cover is ≥75 mm in accordance with Clause 3.3.1.4 of BS 8110: Part 1: 1985. The additional thickness should also be ignored for the purpose of crack width calculation.

If APM4 is to be adopted and blinding concrete is to be used as part of the APM, the blinding concrete should be of the same quality as the foundation construction.

In general, it should be realised that some attack of this sacrificial concrete can be expected. Caution should be exercised in the use of this APM if such attack could affect the structural integrity, for example by introduction of expansive forces or the reduction of frictional forces.

5.7 Addressing site drainage - APM5

The concept of this measure is to consider routes by which aggressive groundwater can reach below-ground concrete and, where necessary, to modify the site drainage to minimise contact between the groundwater and concrete.

For all sites, the engineer should consider the implication of the proposed development on the ground and surface water regimes. If 'addressing site drainage' is being utilised as an APM, the engineer will need to carry out a detailed assessment of water movements before (see Part 1, Section 4) and after construction. As indicated below, there are various options available to reduce the risk of aggressive groundwater coming into contact with buried concrete including, 'deemed to satisfy', redesigning the structure to avoid the drainage problems, and the construction of cut-off barriers and cut-off drains. Care is needed during construction to avoid temporary or permanent situations which increase risks. Drains should be inspected and maintained to avoid leakage close to buried concrete.

There will generally be three groundwater/ concrete environments to be considered in respect of addressing drainage as an APM:

- After construction, the concrete will be surrounded by relatively impermeable ground, such as undisturbed clay strata, through which there is little or no movement of groundwater. In this situation, the APM relating to site drainage is deemed to be already satisfied for concrete, provided it is not subject to a hydrostatic gradient from groundwater of greater than five times the thickness of the concrete. In particular, a consideration of ground water pressures will be needed for structures such as basements and retaining walls that have one side exposed to air.
- It is initially intended that naturally impermeable ground surrounding the concrete be cut through, for example by excavation for construction access or trenches for service pipes, allowing access to groundwater from more permeable ground and/or to surface water. In this situation, the recommended APM can sometimes be achieved by redesigning the works so that the concrete remains surrounded by impermeable ground which forms a barrier to movement of aggressive groundwater. An example might be the use of a piled foundation or trenchfill foundation for a structure, rather than a spread footing constructed in an open excavation. If breaching the naturally impermeable ground around concrete is unavoidable, the APM can often be achieved by resealing the possible routes by which groundwater can reach the concrete. Alternatively, it can be achieved by designing site drainage which will conduct the groundwater in trenches and excavations away from the concrete, rather than towards it. As noted in Part 1, Section 4, it is particularly important in aggressive ground conditions to avoid the situation where a backfilled excavation acts as a sump, ponding water against the structure. This would be particularly aggressive to concrete if the backfill contains sulfate or sulfide-bearing material, for example pyritic clay.
- The concrete will be surrounded by relatively permeable ground. In such cases, the recommended APM can be achieved by installing site drainage designed to remove any aggressive groundwater from the vicinity the concrete and to conduct it safely away. The local authority and/or the Environment Agency may need to be consulted to ensure that any change in the drainage does not adversely affect surrounding land and groundwater.

Particular care will need to be taken with site drainage if a source of highly mobile groundwater is identified on the site. Large housing developments and civil engineering undertakings, particularly works involving road and bridge construction, usually involve considerable disruption to the natural drainage. It is standard procedure to provide for identified water courses to be accommodated in the new site layout or for them to be efficiently diverted. However, in permeable ground, particularly on or adjacent to slopes, many minor water channels may exist that could be a source of aggressive highly mobile water, albeit intermittently. Construction works themselves, particularly trenches for the access of services to buildings, may create further pathways for flows of highly mobile water.

6. Contract documentation

Clients, designers and specifiers should ensure that the recommendations of this Special Digest are included in contract documents.

The preferred approach is for the specifier to provide sufficient information to allow a contractor and concrete supplier to offer 'a package' of proposals to comply with the recommendations, as it could provide the basis for alternative specifications being offered that may reduce construction costs.

Such a information should include as a minimum for each structure:

- Structural Performance Level (SPL).
- DS Class of the ground.
- Aggressive Chemical Environment for Concrete (ACEC) Class of the ground.
- Concrete grade (strength).
- Inclusion in the construction design of any details which could be regarded as Additional Protective Measures, for example drainage, permanent surface protection to concrete.
- Other concrete restrictions.
- Other site constraints.

Some contracts may alternatively opt for full prescription of the concrete requirements and APM. In this case the contact documentation should contain:

- DS-Class of the ground.
- ACEC classification of the ground.
- Design Chemical Class (DC Class) of concrete.
- Any restriction on Aggregate Carbonate Range.
- Any restriction on cement or combination group.
- Required concrete strength.
- Number and (optionally) type of Additional Protective Measures (APM) required.
- Other concrete restrictions.
- Other site constraints.

The project specification should clearly state whether any APM are needed that are not shown on the Contract drawings and whether any particular types are required or preferred.

The contract between the contractor and the concrete producer should always include:

- DC Class of concrete (specifying the appropriate DC Class determines what ACR the producer may use).
- Maximum nominal aggregate size.
- Workability.

The contract may include:

- Grade of concrete.
- Any further restrictions on cement or combination group.
- Any other requirements.

7. British and European standards

Many of the recommendations in the previous Digest 363 which dealt with sulfate and acid attack of concrete in the ground were also embodied in British Standards, notably in BS 5328: *Concrete*. Amendments to BS 5328 will be issued following publication of this Special Digest to bring the two documents into line. It is, however, intended to replace BS 5328 by the new European Standard BS EN 206-1: Concrete and its complementary British Standard, BS 8500, on 1 December 2003.

The classification in BS EN 206-1 for chemical aggressiveness is limited to natural soils and Static water conditions. For this reason, BS 8500 is following this Special Digest with its far wider range of application. The only proposed difference between the BS 8500 provisions and those for BS 5328 is the classification of cements. BS 8500 uses the BS EN 197-1 cements (see Appendix 2, Table A2.1 on page 15) plus residual British Standards for cements not covered by BS EN 197-1. Small adjustments to the limits on composition have also been proposed to facilitate specification, for example any CEM III/B cement will be regarded as being sulfate-resisting.

The British Standard nomenclature has been retained for the various types of cement used in this Special Digest. This will be replaced by European Standards within the next few years when the cement notation in BS EN 197-1 will be introduced. After the British Standard for a particular cement has been withdrawn, concrete specified using the superseded BS notations will contain the equivalent BS EN cement. In practice, this means that the cement will be the same except for its title.

Appendix 1: Expert use of special cements

The expert use of special cements, such as supersulfated cement to BS 4248 and calcium aluminate cement to BS 915 can produce concretes with very good chemical resistance. Supersulfated cement is not currently produced in the UK, but in high quality concrete it has good sulfate resistance and a good reputation for acid resistance provided particular care is taken in the surface curing. Current research on the durability of calcium aluminate cement concrete indicates that their high sulfate and acid resistance is due in part to the formation of a resistant surface zone. Close control must be maintained over the mix proportions, temperature, curing conditions and free water/cement ratio, or there will be a high risk that conversion could reduce the strength and chemical resistance of the concrete. A minimum cement content of 400 kg/m^3 and a total water/cement ratio of not more than 0.40 should be used. Additionally, preventing the surface of the concrete drying out during the first day of curing will ensure continuing hydration and help to maintain the protective surface zone.

Calcium aluminate cements are not covered by BS 8110, but recent revisions to the Building Regulations Approved Documents do not preclude their use in structural concrete provided long-term properties are adequate for purpose and can be reliably predicted.

Appendix 2: Reclassification of cement groups – preliminary table

The British Standard nomenclature for cements is used throughout this Special Digest. The European Standard for common cements is BS EN 197-1 in which Portland cement is known as CEM I cement. BS EN 197-1 does not cover cements specifically designed to be used as sulfate-resisting cements. These are intended to be covered in a separate part of the Standard but a draft has yet to be agreed.

At the time of publication of this Special Digest, a new British Standard, BS 8500: *Concrete* is in preparation. This is complementary to BS EN 206-1. As most British Standards for cement will be withdrawn in April 2002, BS 8500 uses the European classification for cements (as given in BS EN 197-1). To simplify specification, some minor adjustments to the limits for slag content have been introduced to avoid sub-division of sub-types. A comparison between current British Standards and the notation used in BS EN 197-1 is given in Table A2.1.

Table A2.1 Comparison between British Standards current or to be withdrawn and notation for common cements in BS EN 197-1

British Standard: To be withdrawn on 1 April 2002	British Standard: To co-exist beyond 1 April 2000	Cement	BS EN 197-1 notation	Clinker content (%)	Content of other main constituents (%)
	BS 4027: 1996	Sulfate-resisting Portland cement	-	100	-
BS 12: 1996		Portland cement	CEM I	95 to 100	-
	BS 146: 1996 [1]	Portland-slag cement	CEM II/A-S	80 to 94	6 to 20
			CEM II/B-S	65 to 79	21 to 35
		Portland-silica fume cement	CEM II/A-D	90 to 94	6 to 10
		Portland-pozzolanic cement	CEM II/A-P	80 to 94	6 to 20
			CEM II/B-P	65 to 79	21 to 35
			CEM II/A-Q	80 to 94	6 to 20
			CEM II/B-Q	65 to 79	21 to 35
BS 6588: 1996		Portland-fly ash cement	CEM II/A-V	80 to 94	6 to 20
			CEM II/B-V	65 to 79	21 to 35
			CEM II/A-W	80 to 94	6 to 20
			CEM II/B-W	65 to 79	21 to 35
		Portland-burnt shale cement	CEM II/A-T	80 to 94	6 to 20
			CEM II/B-T	65 to 79	21 to 35
		Portland-limestone cement	CEM II/A-L	80 to 94	6 to 20
BS 7583: 1996			CEM II/A-LL	80 to 94	6 to 20
			CEM II/B-L	65 to 79	21 to 35
			CEM II/B-LL	65 to 79	21 to 35
		Portland-composite cement	CEM II/A-M	80 to 94	6 to 20
			CEM II/B-M	65 to 79	21 to 35
	BS 146: 1996 [1]	Blastfurnace cement	CEM III/A	35 to 64	36 to 65
BS 4246: 1996 [2]			CEM III/B	20 to 34	66 to 80
			CEM III/C	5 to 19	81 to 95
		Pozzolanic cement	CEM IV/A	65 to 89	11 to 35
	BS 6610: 1996 [3]		CEM IV/B	45 to 64	36 to 55
		Composite cement	CEM V/A	40 to 64	36 to 60
			CEM V/B	20 to 38	62 to 80

Notes

[1] BS 146: 1996 is to be revised to cover only low early strength classes for blastfurnace slag cements outside the scope of BS EN 197-1 with granulated blastfurnace slag content greater than 35%

[2] BS 4246: 1996 is to be incorporated into the revised BS 146. Following this revision 'High slag blastfurnace slag cement' conforming to BS 146 will be equivalent to cement previously conforming to BS 4246:1996

[3] BS 6610:1996 covers a strength class of 22.5N which is outside the scope of BS EN 197-1

Appendix 3: Determining Aggregate Carbonation Range

A3.1 Procedures for determining Aggregate Carbonate Range (ACR)

Three Aggregate Carbonate Ranges A, B and C have been defined for use in this Special Digest. These are expressed in terms of '$CaCO_3$ equivalent' as a percentage of the total mass of aggregate. The term 'calcium carbonate equivalent' is used since the amount of carbon dioxide (CO_2) released from the aggregate when treated with acid is used as the defining measure. Multiplying the percentage amount of CO_2 by 100/44 gives the calcium carbonate equivalent content. The presence of dolomitic limestones or dolomite (magnesium carbonate) in the aggregate is also covered by this procedure. Such material leads to a slight overestimate of the amount of carbonate present in the aggregate, the maximum $CaCO_3$ equivalent being 108%. However, this is on the safe side for concrete specification.

The limits for ACR A, B and C for cases where all of the carbonate was in, firstly, the coarse and, secondly, the fine aggregate are indicated in Section 2.2 and Table 4. A determination of ACR for any given distribution of carbonate in an aggregate combination may be obtained by one of the following three procedures, (i) to (iii). Because aggregate fractions may derive from different sources, it is generally the concrete producer who determines the ACR of an aggregate combination using data from the aggregate producers on the percentage of $CaCO_3$ equivalent in the coarse and fine aggregates.

(i) Using the chart in Figure A2.1
The position of an aggregate combination on the chart can be calculated from the measured carbonate content of the coarse and fine aggregates as follows:

a) Prepare test portions of separate fine *(f)* and coarse *(c)* aggregates, as detailed in Clause 5 of BS 812: Part 102.

b) Determine the percentage of carbon dioxide (% CO_2) content of each test portion by a suitable method (for example Clause 5 of BS EN 196-21:1992).

c) Convert each concentration of CO_2 to $CaCO_3$ equivalent by multiplying by 100/44 to obtain values for *Pcf* and *Pcc* where:
Pcf = % $CaCO_3$ equivalent in the fine aggregate,
Pcc = % $CaCO_3$ equivalent in the coarse aggregate.

d) Calculate the % of $CaCO_3$ equivalent X in the fine aggregate as a % of the total $CaCO_3$ equivalent in the combined aggregates to give the x-axis value on the chart, using:

$$X = 100 \times \left(\frac{Pcf}{100} \times F\right) \div \left\{\left(\frac{Pcf}{100} \times F\right) + \left(\frac{Pcc}{100} \times C\right)\right\}$$

where F and C are the masses (kg/m^3) of the fine and coarse aggregates respectively.

e) Calculate total $CaCO_3$ equivalent T as a % of the total mass of aggregates to give the y-axis value on the chart, using:

$$T = \gamma \times 100 \times \left\{\left(\frac{Pcf}{100} \times F\right) + \left(\frac{Pcc}{100} \times C\right)\right\} \div (F + C)$$

where γ is a safety factor that allows for source and testing variability. The Quarry Products Association (QPA) [3] suggest using a value for γ of 1.15.
Note: Total $CaCO_3$ equivalent may be more than 100% owing to use of safety factor (and/or owing to the presence of dolomite.

f) Identify applicable Aggregate Carbonate Range A, B or C from x-axis and y-axis values using the chart. A worked example (A) is given in the box on page 17.

(ii) Using linear equations
The final step in the ACR classification set out in Figure A2.1 can be carried out numerically using the linear equations based on guidance from the QPA [3]:

- Test calculation (a) - if

$$T > 30 - 20 \times \left\{\left(\frac{Pcf}{100} \times F\right) \div \left(\left(\frac{Pcf}{100} \times F\right) + \left(\frac{Pcc}{100} \times C\right)\right)\right\}$$

then the aggregate combination is Range A.

- Test calculation (b) - if

$$T > 12 - 10 \times \left\{\left(\frac{Pcf}{100} \times F\right) \div \left(\left(\frac{Pcf}{100} \times F\right) + \left(\frac{Pcc}{100} \times C\right)\right)\right\}$$

and

$$T \le 30 - 20 \times \left\{\left(\frac{Pcf}{100} \times F\right) \div \left(\left(\frac{Pcf}{100} \times F\right) + \left(\frac{Pcc}{100} \times C\right)\right)\right\}$$

then the aggregate combination is Range B;

- Test calculation (c) - if

$$T \le 12 - 10 \times \left\{\left(\frac{Pcf}{100} \times F\right) \div \left(\left(\frac{Pcf}{100} \times F\right) + \left(\frac{Pcc}{100} \times C\right)\right)\right\}$$

then the aggregate combination is Range C.

A worked example (B) is given in the box on page17.

(iii) Use of petrographic procedures
Applicable carbonate contents can also be determined by petrographic procedures as optionally directed by Amendment No 1: 2000 of BS 882:1992, using the procedures described in BS 812: Part 104: 1994.

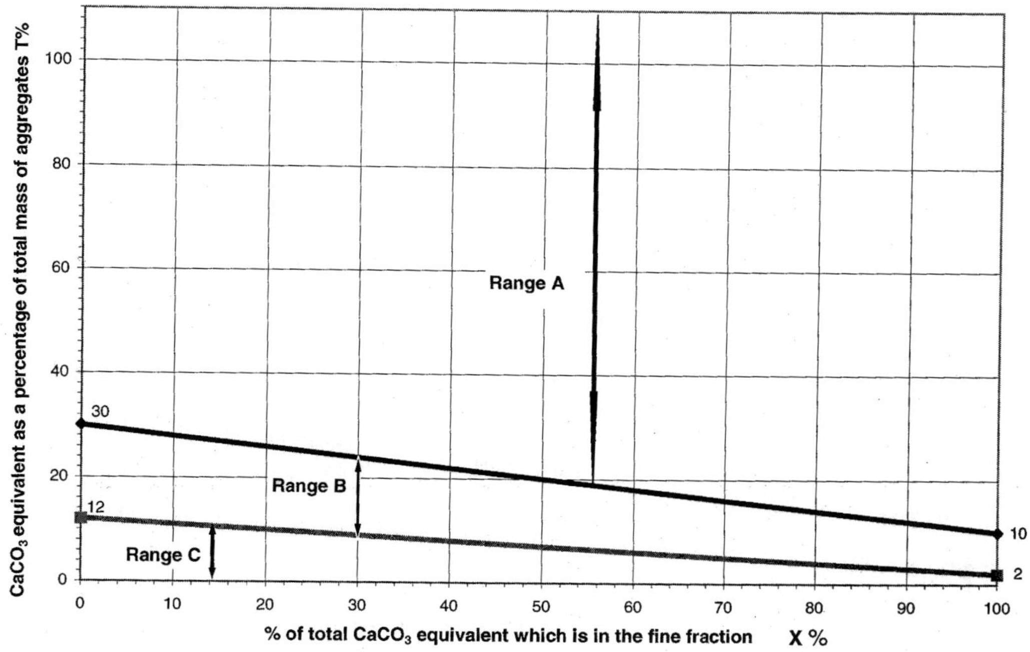

Figure A2.1 Aggregate carbonate range determination for specification of concrete

A3.2 Declaration of aggregate combination as Range A

The procedures in A3.1 provide for a determination of ACR to an accuracy sufficient for all concrete specification conforming with this Special Digest. However, it is important to note that in practice such precise determination is not always used for classifying Range A aggregates. Such practice is endorsed by the guide [3], which states that 'The producer may declare any aggregate combination as being Range A without testing'.

The justifications for this approach to aggregate classification are:

- Some sources of aggregates are so obviously Range A, eg crushed limestone, that there is generally no point in carrying out Range classification tests.
- The ACR of some aggregate sources is difficult and/or costly to classify due to inherent variability.
- It is apparent from Table 6, that for all except one concrete specification (DC-4m), Range B and C aggregates can always be used as a direct replacement for Range A.

As Note [6] to Table 6 indicates, the use of Range A aggregates classified by such above declaration is not permissible when Group 2b cements are to be used in DC-4m concrete. For this exceptional case only, Range A aggregate shall be used which has been authoritatively classified on the basis of the carbonate content, determined by either chemical or petrographic procedures. This is because the research evidence to date indicates that slag cements will only give sufficient protection against sulfate attack in DC-4m conditions if substantial carbonate is present.

Worked Example A

Consider concrete mix with:

- 350 kg/m^3 cement;
- 700 kg/m^3 fine aggregate of which 25% is calcium carbonate (limestone);
- 1400 kg/m^3 coarse aggregate of which 50% is calcium carbonate (limestone).

Then:

F= 700 kg/m^3, C= 1400 kg/m^3, Pcf = 25%, Pcc = 50%

Therefore from d): x-axis value on the chart is given by:

$$X = 100 \times (Pcf/100 \times F) \,/\, \{(Pcf/100 \times F) + (Pcc/100 \times C)\}$$

$$= \frac{100 \times (25/100 \times 700)}{(25/100 \times 700) + (50/100 \times 1400)}$$

$$= \frac{100 \times 175}{175 + 700}$$

$$= 20\%$$

from e): y-axis value on the chart is given by:

$$T = \gamma \times 100 \times \{(Pcf/100 \times F) + (Pcc/100 \times C)\} / (F + C)$$

$$= \gamma \times 100 \times \frac{(25/100 \times 700 + 50/100 \times 1400)}{700 + 1400}$$

$$= \gamma \times 100 \times \frac{(175 + 700)}{2100}$$

$$= \gamma \times 42\% = 48\% \text{ for QPA suggested } \gamma = 1.15$$

X and T plot on Figure A2.1 as Range A aggregate.

Worked Example B

Alternatively, using:

Test calculation (a) in (ii):

$$30 - 20 \{ (Pcf/100 \times F) / [(Pcf/100 \times F) + (Pcc/100 \times C)]$$

$$= 30 - \frac{20 \times (25/100 \times 700)}{(25/100 \times 700) + (50/100 \times 1400)}$$

$$= 30 - \frac{20 \times 175}{175 + 700}$$

$$= 30 - 4$$

$$= 26$$

Since, from previous box, T = 48, the condition:

$T > 30 - 20 \{ (Pcf/100 \times F) / [(Pcf/100 \times F) + (Pcc/100 \times C)] \}$

is satisfied and the aggregate combination classifies as Range A.

Appendix 4: Comparison with Digest 363

In Table 1 of the previous Digest 363, specifications for concrete were given for ground conditions defined solely by (basic) Sulfate Class, and also for one (basic) type of concrete construction and exposure condition (cast-in-situ concrete 140 mm to 450 mm thick, well compacted and not carbonated or subject to hydrostatic pressure). Other exposure conditions and types of concrete were dealt with by modifications to the basic classification given in subsidiary Tables 1a and 1b.

Table A4.1 gives a basic correlation between Sulfate Class as used in Digest 363 and the Aggressive Chemical Environment for Concrete (ACEC) Class and Design Concrete (DC) Class as used in this Special Digest. The correlation between Sulfate Class and ACEC Class is for the basic case of natural ground and non-acid (pH>5.5) mobile groundwater. By definition, in Table A4.1 the 'basic' DC Classes are 'at-par' with the ACEC Classes and basic Sulfate Classes.

In practice, the ACEC Classes derived in Table 2 of this Special Digest departs from the at-par correlation given in Table A4.1 to cater also for brownfield sites, a range of pH conditions and the 'Static' groundwater condition.

Similarly, the at-par correlation between ACEC Class and DC Class in Table A4.1 is expanded in Table 7 of this Special Digest to give a range of specifications for all concrete to be used in aggressive ground together with the number and type of APM (see Section 4.2.3). Appropriate DC Classes are recommended for each of the three Structural Performance Levels (see Part 2, Section 3) and, within these, either two or three concrete thickness categories. These recommendations are based on the principles set out in Section 4.2.2, which have evolved from the rules given in Digest 363 taking into account the finding of the Thaumasite Expert Group [1], but the detailed recommendations are seasoned with some additional engineering judgement.

Appendix 5: Glossary of terms

Aggregate Carbonate Range (ACR)

A new system of classifying aggregates that takes account of recent occurrences of TSA in below-ground concrete containing carbonate aggregates. Three Ranges of aggregate combinations have accordingly been defined, denoted A, B and C. An indication of the permitted amounts of calcium carbonate in these is given in Table 4. See Part 2, Section 2.2 and Appendix 3.

Aggressive Chemical Environment for Concrete classification (ACEC Class)

A new system for the classification of aggressive ground conditions for concrete. It is derived from Design Sulfate Class, taking additional account of the type of site (natural or brownfield) and the mobility and pH of groundwater. Brownfield, 'Mobile' water and low pH are all adverse conditions, which lead to a designation of a more severe ACEC class.

Additional Protective Measures (APM)

APM are defined in this Special Digest as extra measures taken to protect concrete where it is considered that the basic provisions of the concrete specification might not provide adequate resistance to chemical attack (and in particular to TSA) for some uses of concrete. Five Options for APM are listed in Table 8. The number of APM required is greater for highly aggressive ground conditions and/or a higher required Structural Performance Level. See Part 2, Section 5.

Brownfield sites

For the purposes of this Special Digest, a brownfield site is one that has been subject to former industrial development or waste deposition, and which may contain aggressive chemicals in surface waste accumulations or in ground penetrated by leachates. See Part 1, Section 6.1.2.

Table A4.1 Basic correlation between Sulfate Classes of Digest 363 and ACEC Class and DC Classes of this Special Digest

Sulfate Class (Digest 363)	Magnesium content	Type of site	pH	Water mobility	ACEC Class (Table 2)	Design Chemical Class [1] (Table 6)
1		Natural	>5.5	Mobile	AC-1	DC-1
2		Natural	>5.5	Mobile	AC-2	DC-2
3		Natural	>5.5	Mobile	AC-3	DC-3
4	Low	Natural	>5.5	Mobile	AC-4	DC4
4	High	Natural	>5.5	Mobile	AC-4m	DC4m
5	Low	Natural	>5.5	Mobile	AC-5	DC4
5	High	Natural	>5.5	Mobile	AC-5m	DC-4m

Note

[1] For general use of cast-in-situ concrete greater than 140 mm and less than 450 mm thick.
Concrete well compacted with no face exposed to air.

Design Sulfate Class (DS Class)
A five-level classification for sites based on the sulfate (including Potential sulfate) content of the ground and/or groundwater. The sulfate classification (see Part 1, Section 5.1 and Figures 1-6 to 1-8) is dependent on the type of site and the presence or absence of substances including magnesium ions, pyrite and, for pH less than 5.5, chloride and nitrate ions. See Part 1, Section 6.2.

Design Chemical Class (DC Class)
A new system for classifying concrete that is required to be resistant to chemical attack. The DC Class is derived from the ACEC Class and takes into account a number of factors, including the type of concrete element, the method of curing, its mode of exposure to the aggressive ground, and the required Structural Performance Level. For a given DC Class, options are given in Table 6 for concrete specifications using particular aggregate combinations (of specified ACR) and types of cement.

Enhanced concrete quality
An increase in concrete quality used as an Additional Protective Measure (APM). The necessary enhancement may be determined from Table 6. In this Table, bold horizontal lines separate the various concrete qualities in respect of aggressive ground. Use of the option 'Enhance concrete quality' as an APM can be satisfied by using the recommendations of the next higher DC Class.

Mobile groundwater
Mobile groundwater sites include those where water is held in pores and structural discontinuities in the soil and which is free to flow into an excavation to give a standing water level. The ground permeability will generally be greater than 10^{-6} m/s. Ponds or similar accumulations of free water are regarded as mobile. See Part 1, Section 4.

Oxidisable Sulfides (OS)
This is the amount of sulfate that may result from the oxidation of pyrite in the ground (most likely due to ground disturbance).
OS (expressed as % SO_4) can be calculated from:

$$OS = PS - AS$$

where:

PS = Potential Sulfate content as % SO_4
AS = acid-soluble sulfates content as % SO_4

See Part 1, Section 6.1.3.

Pyritic ground
Ground (natural or fill) that contains sulfides, such as pyrite. It is essential to take account of the 'potential' sulfate content, which might result from the oxidation of such ground following ground disturbance.

Sacrificial layer
An Additional Protective Measure for concrete which comprises increasing the thickness of a construction element so that aggressive chemicals to which it is exposed are absorbed in a sacrificial outer layer. Use of this measure will not be appropriate in circumstances where the surface of the concrete must remain sound to prevent loss of frictional resistance or settlement, for example for skin friction piles.

Static groundwater
Static groundwater sites will be confined to sites where the ground is either permanently dry or contains water but is relatively impermeable (virtually no water movement is possible). The mass permeability in the latter case will generally be less than 10^{-6} m/s. A typical example would be clayey soils with tight fissures and no included sand or silt horizons. See Part 1, Section 4.

Structural Performance Level (SPL)
The concept of Structural Performance Level has been introduced to take account of the structural performance factors, such as the consequence of serious erosion and ease of repair, in the specification of concrete and associated APM. Three Structural Performance Levels have been defined: Low, Normal and High. See Part 2, Section 3 and Table 5.

Thaumasite form of sulfate attack (TSA)
A form of sulfate attack that typically affects concrete containing a source of calcium silicate (mostly derived from cementitious calcium silicate phases present in Portland cements) and carbonate (eg calcium carbonate, as limestone, in coarse and/or fine concrete aggregates). The attack requires a source of sulfates (generally from sulfates or sulfides in the ground), the presence of mobile groundwater and low temperature (TSA is most active below 15°C). See Part 1, Section 3.1.2.

Total Potential Sulfate (TPS)
Total potential sulfate content which might result from the combination of sulfates already present in the ground, plus that which may be added due to oxidation of pyritic ground. It is calculated from the stoichiometric equation:

$$TPS = 3.0 \times TS$$

where:

TPS = Potential Sulfate content as % SO_4
TS = Total sulfur content as % S.

See Part 1, Section 6.1.3.

APPENDIX 6: Design guides for different categories of buildings and structures in Parts 3 and 4

Category of building/ structure	Design guide
Part 3: Foundations for non-domestic buildings including any residential building of over three storeys above ground level	1a: Cast-in-situ concrete greater than 140 mm and less than 450 mm thick.
	1b: Cast-in-situ concrete greater than 450 mm thick
	1c: Cast-in-situ concrete less than 140 mm thick
	1d: Cast-in-situ concrete greater than 140 mm and less than 450 mm thick, with a hydrostatic head greater than five times the wall thickness
	1e: Cast-in-situ surface-carbonated concrete greater than 140 mm and less than 450 mm thick, with a hydrostatic head greater than five times the wall thickness
	1f: Cast-in-situ concrete piles greater than 140 mm diameter
	1g: Precast ground beams, wall units and piles - normally cured
	1h: Precast ground beams, wall units and piles surface-carbonated by exposure to air
Part 3: Foundations for low-rise domestic buildings	2a: Concrete trench fill and strip foundations
	2b: Concrete raft foundations
	2c. Concrete floor slabs and oversite concrete on fill or hardcore containing sulfates
Part3: Foundation elements for transport structures	3a: Cast-in-situ concrete greater than 140 mm and less than 450 mm thick
	3b: Cast-in-situ concrete greater than 450 mm thick
	3c: Retaining walls with a hydrostatic head greater than five times the wall thickness
	3d: Retaining walls with a hydrostatic head greater than five times the wall thickness, and concrete surface-carbonated by exposure to air
	3e: Cast-in-situ concrete used for piles greater than 140 mm diameter
	3f: Precast ground beams, wall units and piles with concrete normally cured
	3g: Precast ground beams, wall units and piles with concrete normally cured and surface-carbonated
Part 4: Pipeline systems	4a: Concrete pipes and associated units for water and sewer services in natural ground – external surface
	4b: Concrete pipes and associated units for water and sewer services in brownfield sites - external surface
	4c: Concrete pipes and associated units for water and sewer services – internal surface
Part 4: Precast linings for tunnels and shafts	5a: Precast segmental linings and associated units for tunnels and shafts in natural ground for water and sewer services, storage and transportation – external surface
	5b: Precast segmental linings and associated units for tunnels and shafts in brownfield sites for water and sewer services, storage and transportation – external surface
	5c: Precast segmental linings and associated units for tunnels and shafts for water and sewer services – internal surface
Part 4: Precast box culverts	6a: Precast concrete box culverts in natural ground – external surface
	6b: Precast concrete box culverts in brownfield sites – external surface
	6c: Precast concrete box culverts for water and sewage services – internal surface
Part 4: Precast concrete masonry blocks	7a: Precast aggregate concrete blocks used below ground
	7b: Autoclaved aerated concrete (Aircrete) blocks used below ground

References

[1] Department of Environment, Transport and the Regions. *The thaumasite form of sulfate attack: Risks, diagnosis, remedial works and guidance on new construction.* Report of the Thaumasite Expert Group. DETR, January 1999.
[2] Price, WF. *Controlled permeability formwork.* CIRIA Report C511, 2000.
[3] Quarry Products Association. *Practical application of the DETR Expert Group Report on the thaumasite form of sulfate attack on buried concrete.* London, 1999.

British Standards Institution
BS 12: 1996 Specification for Portland cement
BS 146: 1996 Specification for Portland blastfurnace cements
BS 812: Testing aggregates
Part 102: 1989: Methods for sampling
Part 104: 1994: Method for qualitative and quantitative petrographic examination of aggregates
BS 882: 1992 Specification for aggregates from natural sources for concrete
BS 915: Part 2: 1972 Specification for high alumina cement
BS 1047: 1983 Specification for air-cooled blastfurnace slag aggregate for use in construction
BS 3892: Pulverized-fuel ash
Part 1: 1997 Specification for pulverized-fuel ash for use with Portland cement.
BS 4027: 1996 Specification for sulfate-resisting Portland cement
BS 4246: 1996 Specification for high slag blastfurnace cement
BS 4248: 1974 Specification for supersulfated cement
BS 5328: Concrete
Part 1: 1997 Guide to specifying concrete
Part 2: 1997 Methods for specifying concrete mixes
BS 5628: Code of practice for use of masonry
Part 3:1985 Materials and components, design and workmanship
BS 5911: Precast concrete pipes, fittings and ancillary products
Part 100: 1988 Specification for unreinforced and reinforced pipes and fittings with flexible joints
Part 120: 1989 Specification for reinforced jacking pipes with flexible joints
Part 200: 1994 Specification for unreinforced and reinforced manholes and soakaways of circular cross section
BS 6073: Precast concrete masonry units
Part 1: 1981 Specification for precast concrete masonry units
BS 6588: 1996 Specification for Portland pulverized-fuel ash cements
BS 6610: 1996 Specification for pozzolanic pulverized-fuel ash cement
BS 6699: 1992 Specification for ground granulated blastfurnace slag for use with Portland cement
BS 7583: 1996 Specification for Portland limestone cement
BS 8500: Concrete – complementary British Standard to BS EN 206:1 *(in preparation)*
BS 8110: Structural use of concrete
Part 1: 1985 Code of practice for design and construction
BS EN 196-21: 1992 Determination of the chloride, carbon dioxide and alkali content of cement. Clause 5
BS EN 197-1: 2000 Cement
Part 1: Composition, specification and conformity criteria for common cements
BS EN 206-1: 2000 Concrete
Part 1: Specification, performance, production and conformity